Edward Jarvis

Influence of Distance from and Nearness to an Insane

Hospital

Edward Jarvis

Influence of Distance from and Nearness to an Insane Hospital

ISBN/EAN: 9783337375591

Printed in Europe, USA, Canada, Australia, Japan

Cover: Foto ©berggeist007 / pixelio.de

More available books at **www.hansebooks.com**

INFLUENCE OF DISTANCE

FROM AND NEARNESS TO

AN INSANE HOSPITAL

ON ITS USE BY THE PEOPLE.

BY EDWARD JARVIS. M. D.

INFLUENCE OF DISTANCE FROM AND NEARNESS TO AN INSANE HOSPITAL ON ITS USE BY THE PEOPLE.

BY EDWARD JARVIS, M. D.

At the present moment, when the Legislature of New York is proposing to establish another large central hospital for the care of the insane of the whole State, making it necessary for the people of every county. however near or remote, to send their curable patients to one institution, at Utica, and all their incurable patients to another central institution, it is worth while to examine the history of the past and see what has been the effect of this endeavor to concentrate all the lunatics of the State in one place, and how far the blessings of such an institution, its privileges and advantages have been practically given to, and enjoyed by the people in the various parts of the State.

At the same time it will be well to examine the history of similar institutions in other States, and see how far they have accomplished the whole purpose of their creation, in healing or caring for the mental maladies of their people, in all their near and remote districts.

An insane hospital is, and must be, to a certain extent, a local institution. People will avail themselves of its privileges in some proportion to their nearness to it. No liberality of admission, no excellence of its management, no power of reputation can entirely overcome the obstacle of distance, expense, and of the difficulties of transporting lunatics, or the objection of friends to sending their insane patients far from home, and out of the reach of ready communication.

The operation of this principle, in some degree, seems probable to any one who gives a thought to the matter; but the facts, the particular history of those institutions, in which the records of the homes of their patients are kept, show that the objection of distance prevails with all of them, and that those hospitals have been and are used by those who live near by, much more than by those who live farther off; and consequently they are practically much more local in their usefulness than they are intended or are supposed to be.

The State Hospital at Utica was opened in 1843, and offered to the people of every county, both near and remote, the same conditions. The people of Oneida, Schoharie, Orange, Washington and Chautauqua were alike invited to send their insane, on the same terms. Between them there was and could be no difference of advantage, after their patients should be placed in the hospital; the only difference was in the distances between their homes and the institution, in the labor, cost and burden of travelling to a hospital with a lunatic.

To make this matter more certain and to show the difference of enjoyment to the eye, the whole State has been divided into four Districts, according to their distance from the hospital.

The first District is Oneida county, in which the hospital is situated.

The second District consists of eleven counties : Chenango, Cortland, Fulton, Herkimer, Lewis, Madison, Montgomery, Onondaga, Oswego, Otsego, Schoharie.

These are mostly within 60 miles of Utica.

The third District includes seventeen counties, which are from 60 to 120 miles distant: Albany, Broome, Cayuga, Columbia, Delaware, Greene, Hamilton, Jefferson, Rensselaer, Saratoga, Schenectady, Seneca, Tioga, Tompkins, Warren, Washington, Wayne.

The fourth District includes the most distant counties, which are from 120 to 350 miles from Utica: Allegany, Cattaraugus, Chautauqua, Chemung, Clinton, Dutchess, Erie, Essex, Franklin, Genesee, Livingston, Monroe, Niagara, Ontario, Orange, Orleans, Putnam, Queens, Richmond, Rockland, Schuyler, Steuben, St. Lawrence, Suffolk, Sullivan, Ulster, Westchester, Wyoming, Yates.

These four Districts include all the counties of the State, except New York and Kings, which have each hospitals of their own, and, therefore, little or no occasion or inducement to send patients to Utica.

The population of each of these Districts has been ascertained and calculated for each of the twenty-three years, 1843 to 1865, inclusive, since the hospital was opened. The number of patients sent to the hospital from each District, within that period, has also been ascertained. Taking, then, the sum of the annual populations for twenty-three years, and dividing it by the number of patients sent in that time, shows the proportion of patients which each District has sent out of its whole number of people. These numbers and facts are presented in the following tables:

Annual Population of the Districts of New York.

YEAR.	DISTRICTS.			
	I.	II.	III.	IV.
1843	84,990	411,281	625,224	998,656
1844	84,880	412,350	629,913	1,010,640
1845	84,776	413,445	634,561	1,022,799
1846	87,658	421,217	647,252	1,053,687
1847	90,538	429,135	660,197	1,085,508
1848	93,619	437,202	673,400	1,118,290
1849	96,805	445,421	686,868	1,152,062
1850	99,566	453,768	700,803	1,186,728
1851	100,959	456,490	704,657	1,213,310
1852	102,372	459,228	708,532	1,240,488
1853	103,805	461,983	712,428	1,268,274
1854	105,258	463,754	716,346	1,296,683
1855	107,749	465,291	719,997	1,326,918
1856	107,265	469,013	728,132	1,307,811
1857	106,783	472,765	736,361	1,289,049
1858	106,303	476,547	744,681	1,270,487
1859	105,825	480,349	753,095	1,252,192
1860	105,202	486,212	761,460	1,235,347
1861	104,676	484,994	760,547	1,276,113
1862	104,153	483,783	759,635	1,318,224
1863	103,633	482,574	758,724	1,361,725
1864	103,115	481,368	757,814	1,406,661
1865	102,713	480,236	756,893	1,454,825
Total,	2.292,643	10,528,406	16,337,520	28,146,477

For these twenty-three years, 1843 to 1865, in Oneida county, the sum of the annual populations was 2,292,643 who sent 827 patients, or 1 in 2,772 of this number, to the hospital. In the second District, the sum of the annual populations was 10,528,406, who sent 1,809 patients, or 1 in 5,820 of this number to the hospital. In the third District, the sum of the annual populations was 16,337,520, who sent 2,222 patients, or 1 in 7,351 of their number to the hospital. In the fourth District, the sum of the annual populations was 28,146,477, who sent 2,440 patients, or 1 in 11,535 of their number to the hospital.

Population and Patients of Districts.

	DISTRICTS.			
	I.	II.	III.	IV.
Sum of the Annual Population for 23 years,	2,292,643	10,528,406	16,337,520	28,146,477
Patients sent to the Hospital in 23 years,	827	1,809	2,222	2,440
Average Annual Population,	99,680	457,756	710,327	1,223,760
Average Patients sent to the Hospital,	36	78	96	106
Population to one Patient sent to the Hospital in each year,	2.772	5,820	7,351	11,535

This shows a great disproportion in the uses made of the hospital by the people of the near and of the remote counties.

Taking a basis 1,000 for the extent of the enjoyment of the hospital by the remotest Districts, the proportionate enjoyment of the Districts will be: IV, 1,000; III, 1,568; II, 1,981; I, 4,196.

The advantages of the hospital enjoyed by Oneida county have been more than double those enjoyed by the counties next beyond, but within 60 miles; they are nearly threefold those enjoyed by the counties which are from 60 to 120 miles distant; and more than four times as great as those enjoyed by the people of the counties which are more than 120 miles distant.

It will not be supposed that the insane persons who needed the hospital care or treatment in these Districts were in these proportions. It cannot be supposed that the number of lunatics in Oneida county is twice as great as that in Oswego, Fulton, Schoharie, Herkimer, and the other counties beyond Oneida but within 60 miles; or four times as great as that in counties 120 and more miles from this District.

The State Censuses of 1855 and 1865 show the number of the insane in the several counties of New York.

Arranging these in the Districts herein described, according to their distance from Utica, they were in proportion to the population.

Population to One Lunatic in New York.

District.	1855.	1865.
I.	1,224	1,300
II.	1,525	1,611
III.	1,457	1,396
IV.	1,788	1,904

This diversity of advantage of an insane hospital enjoyed by the people of near and remote Districts, is not an accident, nor a peculiarity of New York alone. It is a general and probably universal principle—a natural and necessary law of nature or of humanity—for in all other States whose hospital records of patients' residence have been obtained, the same law is found to be in operation, and the people send their patients to these institutions in proportion to their nearness.

In twenty-six States, for various periods of years, insane hospitals have been in operation, whose doors are and have been open alike to all of their people. The Reports of most of these institutions state the number which have been sent to them from each county. From the others, copies of the records of facts have been obtained, showing the number which the various parts of the States have contributed to fill the wards of those institutions. In order to determine the extent and application of the law of distance in the use of hospitals, these other States and two of the British Provinces have been examined and analyzed in the same way as New York.

They have been divided into concentric districts, making the county in which the hospital is situated the first, and the contiguous counties the second district, and the others more distant. The populations of these several districts have been calculated and determined for each of the years in which the hospital has been in operation, or in which the records of the residence of the patients were kept and have been obtained, and the comparison made of the proportion of patients to population of the several districts.

It should be here stated, that in making these concentric circular divisions, it has been impossible to make them perfectly regular, with an exactly equal radius from the common centre or equal distance of the inner and outer boundary from the hospital, for the counties are very diversely and irregularly shaped, some of them, as in Maine, being nearly 150 miles long. While then a district may be stated to be within certain specified distances from the hospital, circles drawn upon these radii would, on both sides, exclude some part of the territory that belongs to it, and include some that belongs to its neighbor. Nevertheless, these irregularities of border, or exceptions to the rule, will not militate with the general plan, nor vitiate any calculations made upon, or deductions made from, this analysis of the States and hospital receptions.

Twenty-two States and two British Provinces furnish the conditions requisite for the purpose of this report, and are included in the calculations and statements : Maine, New Hampshire, Vermont, Massachusetts, Rhode Island, Connecticut, New Jersey, Pennsylvania, Maryland, Virginia, North Carolina, Michigan, Ohio, Indiana,

Illinois, Missouri, Kentucky, Tennesee, Mississippi, Louisiana, Canada, Nova Scotia.

MAINE.

The hospital is established by and under the control of the State, and open alike to the people of all its parts. It has been in operation from 1840, and has a record of twenty-six years. The hospital is at Augusta, in the county of Kennebec.

The first District consists of Kennebec county.

The second District includes eight counties contiguous to Kennebec: Androscoggin, Franklin, Knox, Lincoln, Oxford, Sagadahoc, Somerset, Waldo.

The third District includes three counties: Cumberland, Hancock, York.

The fourth District consists of four counties: Aroostook, Penobscot, Piscataquis, Washington.

In the first District, Kennebec county, the sum of the annual populations for twenty-six years was 1,519,860; these sent 536 patients to the hospital; equal to one in 2,835 of the living in each year.

In the second District, the sum of the populations was 5,869,616, or an annual average of 225,754. This District sent 1,135, or an annual average of 43.3, equal to one in 5,171 of the living in each year.

In the third District, the sum of populations was 4,414,348, an annual average of 169,782. These sent 784 patients in twenty-six years, an average of 30.1, or one in 5,630 in each year.

In the fourth District, the sum of the populations was 3,448,294, or an annual average of 133,396. They sent 437 patients, an average of 15.8, or one in 7,890 of the people.

NEW HAMPSHIRE.

The hospital was established in 1842, and has been in operation twenty-three years. It is in Merrimac county, which is the first District.

The second District includes the contiguous counties : Belknap, Grafton, Hillsborough, Rockingham, Sullivan. The third District includes the most remote counties : Carroll, Cheshire, Coos, Stafford.

In the first District, the sum of the annual populations, for twenty-three years, is 966,310, an average of 42,013. These sent 396 patients to the hospital, a yearly average of 17.2, or one in 2,440 of the people.

In the second District the sum of populations was 4,406,569, an annual average of 191,589. These sent 1,270 patients, averaging 55.2, or one in 3,469 of the living, in each year.

In the third District, the sum of the annual populations was 2,229,424, averaging 96,931 yearly. These sent 355 patients to the hospital, or 15.4, one in 6,280, in each year.

MASSACHUSETTS.

The Worcester Hospital in Massachusetts was opened in 1833, and was the only State institution of its class until 1854, when the Taunton Hospital was opened. During this period it was open to all the people of the State, and received patients from all the counties. In the classification Suffolk county is omitted, because it had a hospital of its own for paupers from 1838, and the McLean Asylum received most of its private patients.

In the first District, including Worcester county, the sum of populations through twenty-one years was 2,378,573, or an annual average of 123,122. These

B

sent 1,067 patients to the hospital, averaging 50.6, or one in 2,229 in each year.

In the second District of contiguous counties, including Franklin, Hampden, Hampshire, Middlesex, Norfolk, the sum of the annual populations was 6,133,637, an average of 292,078 in each year. These sent 3,872 patients to the hospital, or 185.3 yearly, which is equal to one in 3,872 of the living, in each year.

In the third District of the remote counties of Barnstable, Berkshire, Bristol, Dukes, Essex, Nantucket, Plymouth, the sum of populations, through twenty-one years, was 6,602,777, a yearly average of 314,418. From these, 1,333 patients went to the hospital; being 63.5, or one in 4,953 yearly.

THE EXPERIENCE OF MCLEAN ASYLUM SIMILAR TO THAT OF WORCESTER.

The McLean Asylum has been nearly fifty years in operation, at Somerville, within three miles of five cities, Boston, Chelsea, Charlestown, Cambridge and Roxbury. Although a corporate institution, it is open to all the people of Massachusetts on equal terms; all are invited to send their patients. Of the 154 patients in the house February 19, 1866, from Massachusetts, 80 were from the five cities above mentioned, 30 from other parts of Middlesex county, 24 from Essex county, 8 from Norfolk, 16 from Plymouth, 3 from Barnstable, 2 from Bristol, and 1 from Worcester county.

Residence of Patients in McLean Asylum.

County or City.	Distance.	Population.	Patients.	Population to One Patient.
Five Cities,.........	0– 3	290,665	80	3,633
Middlesex,..........	0–35	165,106	30	5,503
Essex,.............	2–35	165,611	24	6,900
Norfolk,...........	3–30	81,524	8	10,190
Plymouth,..........	15–45	63,074	6	10,384
Bristol,............	20–60	89,505	2	
Barnstable, ...	45–90	35,489	3	
Worcester,.........	25–65		1	

RHODE ISLAND.

The Butler Hospital has been opened sixteen years—1849 to 1865—and equally open to all the people of the State.

In the first District, Providence county, the sum of annual populations was 1,704,913, an average of 106,557 in each year. From these 551 patients went to the hospital, equal to 34.4, or one in 3,094 of the living, yearly.

In the second District, embracing the rest of the State, the sum of populations was 1,076,997, or 67,312 in each year. These sent 204 patients, or 12.7 yearly, being one in 5,279 of the people.

NEW JERSEY.

The Hospital is a State institution, situated at Trenton, in Mercer county, and has been in operation from 1848 to 1865, eighteen years. It is open to all the people of every county on the same conditions.

The first District is Mercer county.

The second District includes eight counties, from 12 to 35 miles from Trenton; Burlington, Essex, Hunterdon, Middlesex, Monmouth, Ocean, Somerset, Union.

The third District includes twelve counties, which are from 35 to 75 miles distant from the hospital: Atlantic,

Bergen, Camden, Cape May, Cumberland, Gloucester, Hudson, Morris, Passaic, Salem, Sussex, Warren.

In the first District the sum of populations for eighteen years was 615,070, or an average of 34,170. From these 273 patients were sent to the hospital, which is 15.1, or one in 2,253 of the people in each year.

In the second District the sum of populations was 5,204,296, or an annual average of 289,128. From these 1,401 patients went to the hospital, which is an average of 77.8, or one in 3,714 of the living in each year.

In the third District the sum of the populations was 5,255,946, an annual average of 291,997. In the eighteen years, 890 patients went from this District to the hospital, which is 49.4, or one in 5,905 of the people in each year.

PENNSYLVANIA.

For many years the Hospital for the Insane at Philadelphia, and the Friends Asylum at Frankford, six miles from that city, both corporate institutions, and the City Pauper Hospital, had been in operation. Most of the patients belonging to Philadelphia county were and are sent to these institutions, and comparatively few have been sent to the State Hospital at Harrisburgh. Therefore the county of Philadelphia is omitted in these statements in respect to Pennsylvania.

The State Lunatic Hospital was opened October 6, 1851, and was the only State institution for the insane until 1857, when the hospital at Pittsburgh was opened to the insane in the western part of the State.

The calculations for the Harrisburgh Hospital are made for the whole State for this period, 1851, to 1857, and for the middle and eastern parts of the State for the subsequent period.

Harrisburgh is on the border of Dauphin county, and is as near to Cumberland, the contiguous county. There- , fore both of these counties are included in the first district.

The second District includes ten counties within 55 miles : Adams, Franklin, Juniata, Lancaster, Lebanon, Northumberland, Perry, Schuylkill, Snyder, York.

The third District, includes the twenty-two counties next beyond the second District, 55 to 110 miles distant from Harrisburgh : Bedford, Berks, Blair, Cambria, Carbon, Centre, Chester, Clearfield, Clinton, Columbia, Delaware, Fulton, Huntington, Lehigh, Luzerne, Lycoming, Mifflin, Montgomery, Montour, Northampton, Sullivan, Union.

The fourth District includes twenty-nine counties, 110 to 250 miles distant : Armstrong, Beaver, Bradford, Bucks, Butler, Clarion, Crawford, Elk, Erie, Fayette, Forest, Greene, Indiana, Jefferson, Lawrence, Mac-Kean, Mercer, Monroe, Pike, Potter, Somerset, Susquehanna, Tioga, Venango, Warren, Washington, Wayne, Westmoreland, Wyoming.

During the period when the hospital at Harrisburgh was the only State institution for the insane, the sum of the annual populations of the several Districts and of the patients sent, and also the annual averages of these were as follows :

In the first District, the sum of the populations was 418,256, a yearly average of 76,046. Their total of patients sent to the hospital was 69, being 12.5 patients and one in 6,061 of the people in each year.

In the second District the sum of populations was 2,190,973, or an yearly average of 398,358, who sent 203 patients to the hospital, equal to 36.9 each year, or one in 10,793 of the living.

In the third District the sum of the populations was 3,678,752, and number of patients 208, which was an annual average of 668,864 people and 37.9 patients, or one in 17,686 of the living.

In the fourth District the sum of people was 4,227,184, and patients 178, an annual average of 768,578 people and 32.3 patients, or one in 23,748 of the living.

From 1857 to the present date, the Harrisburgh Hospital and the Western Hospital have divided the State; certain counties being assigned, by law, to the eastern and certain others to the western institution.

For this period, 1857 to 1865, the eastern and central portions of the State are divided into four districts in reference to the Harrisburgh Hospital, and the western part into three, with reference to Pittsburgh as a centre.

In the eastern and central parts of the State :

In the first District, during these years, the sum of populations was 800,260, and the annual average 88,917. In the whole period 136 patients went to the hospital, which was equal to 15.1, or one in 5,884 people in each year.

In the second District the sum of annual populations was 4,240,788, an average of 471,198 yearly. These sent 404 patients, which is equal to 45, or one in 10,497 people each year.

In the third District the sum of populations was 6,808,921, an average of 756,546 in each year. They sent 391 patients to the hospital in the whole period, which was equal to 43.4, or one in 17,414 people yearly.

In the fourth District the sum of annual populations was 7,669,080, who sent 143 patients to the hospital. The annual average of population was 852,120, and of patients 15.8, or one in 53,629.

WESTERN PENNSYLVANIA.

Allegany county was the first District.

The second District includes five contiguous counties within 50 miles of Pittsburgh : Armstrong, Beaver, Butler, Washington, Westmoreland.

The third District includes fourteen counties, 50 to 125 miles of Pittsburgh : Cambria, Clarion, Crawford, Elk, Erie, Fayette, Greene, Jefferson, Lawrence, Mac-Kean, Mercer, Somerset, Venango, Warren.

In the first District, the sum of populations was 1,613,403, and the annual average 179,267. From these 442 patients were sent, averaging 49.1, or one in 3,650 people yearly.

In the second District, the sum of annual populations was 1,809,991, and the yearly average 201,110. These sent 171 patients, which was equal to 19, or one in 10,584 people in each year.

In the third District, the sum of annual populations was 3,737,948, or an annual average of 415,327. From these 167 patients went to the hospital, which was equal to 18.5, or one in 22,382.

MICHIGAN.

The Asylum for the Insane was established by the State, at Kalamazoo, Kalamazoo county, and commenced operations in August, 1859, and its privileges were offered to all the people of the State—the near and the remote—to all on the same terms.

The first District is Kalamazoo county.

The second District includes seven contiguous counties, within 35 miles of the asylum : Allegan, Barry, Branch, Calhoun, Cass, St. Joseph, Van Buren.

The third District includes twenty counties, from 35 to 100 miles distant : Berrien, Clinton, Eaton, Gratiot, Hillsdale, Ingham, Ionia, Isabella, Jackson, Kent, Lenawee, Livingston, Mecosta, Montcalm, Muskegon, Newaygo, Oceana, Ottawa, Shiawassee, Washtenaw.

The fourth District includes nineteen counties, from 100 to 150 miles distant from Kalamazoo : Bay, Clare, Genesee, Gladwin, Lake, Lapeer, Macomb, Manistee, Mason, Midland, Missaukee, Monroe, Oakland, Osceola, Roscommon, Saginaw, St. Clair, Tuscola, Wexford.

The fifth District includes twenty-three counties, from 150 to 350 miles distant from Kalamazoo : Alcona, Alpena, Antrim, Cheboygan, Chippewa, Crawford, Delta, Emmet, Grand Traverse, Houghton, Huron, Iosco, Kalkasca, Leelenau, Manitou, Marquette, Michilimackinac, Ogemaw, Ontonagon, Otsego, Presque Isle, Sanilac, Schoolcraft.

During the five and one-half years, of which the record is printed, the asylum received 386 patients. The annual populations of the several districts, and the patients received from them were as follows :

The sum of populations in the first District was 151,794, and the patients sent to the asylum 48, making an annual average of 27,599 people and 8.7 patients, or one in 3,762 of the living.

The population of the second District through these years amounted to 830,623, and the patients sent from these to 90. These show an annual average of 151,022 inhabitants and 16.3 patients, or one in 9,229 of the people.

In the third District the sum of populations was 1,885,265, and the number of patients sent from these 170. The averages of these were 342,775 people and 30.9 patients, or one in 11,089 of the living.

In the fourth District the annual populations amounted to 1,037,211, who sent 73 patients to the asylum. The annual averages were 188,583 people and 13.2 patients, or one in 14,208 persons.

In the fifth District, the populations were 292,195, and from these 5 patients were sent to the asylum. During the period under observation, the average population was 53,126, and the yearly average of patients less than one, being one in 58,439 people.

OHIO.

The State Lunatic Asylum began its operations at Columbus, in Franklin county, November 30, 1839. It was the only institution in the State for the insane, except a local hospital at Cincinnati, until 1855, when the Northern Asylum was opened at Newburgh, for the northern and north-eastern counties, and the Southern Asylum at Dayton, for the western and south-western counties. From November, 1839, to 1855, the Columbus Asylum received patients from all the counties, its privileges being equally offered to all.*

The first District includes Franklin county.

* In this classification, Hamilton county having a hospital, is omitted.

C

The second District includes six contiguous counties, within 40 miles of Columbus: Delaware, Fairfield, Licking, Madison, Pickaway, Union.

The third District includes twenty-six counties next beyond those before mentioned, and from 40 to 75 miles from Columbus: Athens, Champaign, Clark, Clinton, Crawford, Fayette, Greene, Guernsey, Hardin, Highland, Hocking, Jackson, Knox, Logan, Marion, Miami, Montgomery, Morrow, Muskingum, Perry, Pike, Richland, Ross, Shelby, Vinton, Wyandot.

The fourth District includes fifty-four counties, from 75 to 150 miles from the asylum: Adams, Allen, Ashland, Ashtabula, Auglaize, Belmont, Brown, Butler, Carroll, Clermont, Columbiana, Coshocton, Cuyahoga, Darke, Defiance, Erie, Fulton, Gallia, Geauga, Hancock, Harrison, Henry, Holmes, Huron, Jefferson, Lake, Lawrence, Lorain, Lucas, Mahoning, Medina, Meigs, Mercer, Monroe, Morgan, Noble, Ottawa, Paulding, Portage, Preble, Putnam, Sandusky, Scioto, Seneca, Stark, Summit, Trumbull, Tuscarawras, Van Wirt, Warren, Washington, Wayne, Williams, Wood.

During these sixteen years the sum and averages of annual populations in, and of patients sent to the Central Asylum were as follows:

In the first District the sum of populations was 1,145,181, an annual average of 39,489. The number of patients sent was 225, which was 7.7, or one in 5,060 people in each year.

In the second District the sum of populations was 3,805,589, or an annual average of 131,227. These sent 521 patients, or 17.9 in each year, equal to one in 7,304 people.

In the third District the sum of people was 15,003,348, averaging 517,356 in each year. From these 1,281 patients went to the asylum, or 9.6 in each year, being one in 11,712 inhabitants.

In the fourth District the sum of the annual populations was 31,154,619, equal to an average of 1,074,297 in each year. The number of patients who went to the asylum was 1,079, equal to 37.2, or one in 28,873 people yearly.

ILLINOIS.

The State Asylum was opened in Jacksonville, Morgan county, in 1847. The printed reports state the residence of the patients to 1864, eighteen years.

The first District is Morgan county.

The second District includes eight counties, within 40 miles: Brown, Cass, Greene, Macoupin, Menard, Pike, Sangamon, Scott.

The third District includes fifteen counties, 40 to 75 miles distant: Adams, Bond, Calhoun, Christian, Fulton, Hancock, Jersey, Logan, Macon, Madison, Mason, McDonough, Montgomery, Schuyler, Tazewell.

The fourth District includes forty-two counties, 75 to 125 miles from the asylum: Bureau, Champaign, Clay, Clinton, Coles, Cumberland, De Kalb, De Witt, Douglas, Effingham, Fayette, Ford, Henderson, Henry, Jasper, Jefferson, Kendall, Knox, La Salle, Lee, Livingston, MacLean, Marion, Marshall, Mercer, Monroe, Moultrie, Peoria, Perry, Piatt, Putnam, Randolph, Richland, Rock Island, St. Clair, Shelby, Stark, Vermillion, Warren, Washington, Wayne, Woodford.

The fifth District includes thirty-five counties, from 125 to 225 miles distant: Alexander, Boone, Carroll,

Clark, Crawford, Du Page, Edwards, Edgar, Franklin, Gallatin, Grundy, Hamilton, Hardin, Iroquois, Jackson, Joe Daviess, Johnson, Kane, Kankakee, Lake, Lawrence, McHenry, Massac, Ogle, Pope, Pulaski, Saline, Stephenson, Union, Wabash, White, Whitesides, Will, Williamson, Winnebago.

In the first District the sum of annual populations was 337,242, or an average of 18,747, who sent 102 patients, or 5.6 yearly; equal to one in 3,306 of the people.

In the second District the sum of annual populations was 2,052,957, an average of 114,053 in each year. From these 261 went to the asylum, which is 14.5, or one in 7,865 yearly.

In the third District the sum of populations was 3,941,236, a yearly average of 218,957, who sent 423 lunatics to the asylum, equal to 23.5, or one in 9,317 in each year.

In the fourth District the sum of populations was 8,462,974, or 470,137 in each year. From among these 720 patients were sent to the asylum, which was equal to 40, or one in 11,753 annually.

The fifth District had, during the period of the operation of the asylum, a total annual population of 6,655,211, or an average of 391,483. From these 427 patients went to the asylum, which is equal to 25.1, or one in 15,585 people in each year.

MARYLAND.

The Maryland Hospital for the Insane has been established in Baltimore. It is a corporate institution, but is open equally to the patients of all parts of the State.

The records of the residence of the patients from 1850 to 1864, inclusive, have been obtained, and on this period of fifteen years, the following calculations and statements are made :

Baltimore city constitutes the first District.

The second District includes thirteen counties, within 50 miles of Baltimore : Anne Arundel, Baltimore, (country part,) Calvert, Carroll, Cecil, Frederic, Harford, Howard, Kent, Montgomery, Prince George, Queen Ann, Talbot.

The third District includes the eight most remote counties, from 50 to 150 miles distant from the hospital : Allegany, Charles, Caroline, Dorchester, St. Mary's, Somerset, Washington, Worcester.

In the first District the sum of the annual population for fifteen years was 2,989,753, an average of 199,250 for each year. These sent 422 patients to the hospital, being 18.1, or one in 7,034 yearly.

In the second District the sum of populations was 4,383,107, or 292,270 yearly. From these 433 patients went to the hospital, being 28.8, or one in 10,122 of the people in each year.

In the third District the sum of populations was 2,461,482, averaging 164,098 a year. These sent 107 patients to the hospital, which was equal to 7.1, or one in 23,009 of the living annually.

VIRGINIA.

No record is found of the residence of the patients sent to the Eastern Asylum at Williamsburgh.

The patients sent to the Western Asylum at Staunton, Augusta county, are from the western counties, which only are included in the districts. This institution went into operation in 1828, and the people of all the western counties were invited to send their lunatic friends to it. The residence of all patients sent from 1828 to 1859, inclusive, thirty-two years, is recorded in the reports that have been published and obtained.

The first District includes only Augusta county.

The second District includes the nine contiguous counties of Albemarle, Amherst, Bath, Greene, Highland, Nelson, Pendleton, Rockbridge, Rockingham, which are 25 to 45 miles from Staunton.

The third District contains the twenty-nine counties which are in the circle 45 to 90 miles from Staunton : Alleghany, Appomattox, Barbour, Bedford, Botetourt, Buckingham, Campbell, Charlotte, Clark, Craig, Culpepper, Cumberland, Fauquier, Fluvanna, Greenbrier, Hardy, Louisa, Madison, Orange, Page, Pocahontas, Prince Edward, Randolph, Rappahannock, Roanoke, Shenandoah, Upshur, Warren, Webster.

The fourth District includes the thirty-five counties that are from 90 to 135 miles from Staunton : Berkley, Braxton, Calhoun, Clay, Doddridge, Fairfax, Fayette, Floyd, Franklin, Frederic, Giles, Gilmer, Halifax, Hampshire, Harrison, Henry, Jefferson, Lewis, Loudon, Marion, Mercer, Monongalia, Monroe, Montgomery, Morgan, Nicholas, Patrick, Pittsylvania, Preston, Prince William, Pulaski, Raleigh, Ritchie, Roane, Taylor.

The fifth and last District includes all the counties from 135 to 330 miles westward, and as far eastward as the middle line between Staunton and Williamsburgh : Bland, Boone, Brooke, Cabell, Carroll, Grayson, Hancock, Jackson, Kanawha, Lee, Logan, Marshall, Mason, McDowell, Ohio, Pleasants, Putnam, Russell, Scott, Smith, Tazewell, Tyler, Washington, Wayne, Wetzel, Wirt, Wise, Wood, Wyoming, Wythe.

In the course of the thirty-two years, 1828 to 1859, the sum of annual populations of the first District was 695,061, or an average of 21,720 in each year. From these 127 lunatics were sent to the asylum, equal to nearly 6, or one in 5,472 of the people yearly.

The sum of populations of the second District was 3,103,376, or 96,980 in each year. These sent 252 patients; equal to an annual average of nearly 8, or one in 12,314 of the living.

The third District had, in the thirty-two years, a sum of annual populations equal to 8,596,820, or 268,650 in each year. These supplied 399 patients to the asylum, which was equal to an annual average of 12.5, or one in 21,570 of the people.

In the fourth District there were. in the course of this period, 9,162,704 people living, or an annual average of 286,334. From these 218 patients went to the asylum, which was equal to 6.8, or one in 24,433 of the people yearly.

In the fifth District the sum of annual populations was · 5,472,933, an average of 171,029 yearly. This District sent 218 patients, or an average of 6.8 in each year to the asylum, equal to one in 25,105 of the whole people.

NORTH CAROLINA.

The State Asylum was opened at Raleigh, Wake county, in 1856, and offered to the people of every part of the State on equal terms. The residences of the patients are stated in the annual reports, which, from the beginning to 1860, are available for the purposes of this article.

The first District is Wake county.

The second District includes the eight contiguous counties, which are within 50 miles of Raleigh : Chatham, Franklin, Granville, Hamett, Johnson, Moore, Nash, Orange.

The third District includes the thirty-three counties next beyond the last. These are from 50 to 100 miles from the asylum : Alamance, Ansan, Bladen, Cabanas, Caswell, Cumberland, Davidson, Davie, Duplin, Edgecombe, Forsythe, Greene, Guildford, Halifax, Jones, Lenoir, Martin, Montgomery, New Hanover, Northampton, Person, Pitt, Randolph, Richmond, Robeson, Rockingham, Rowan, Sampson, Stanly, Stokes, Warren, Wayne, Wilson.

The fourth District includes twenty-six counties, from 100 to 150 miles from Raleigh : Alexander, Beaufort, Bertie, Brunswick, Carteret, Catawba, Chowan, Columbus, Craven, Gaston, Gates, Hertford, Hyde, Iredell, Lincoln, Mecklenburgh, Onslow, Pasquotank, Perquimans, Piatt, Surrey, Tyrrell, Union, Washington, Wilkes, Yadkin.

The fifth District includes the eighteen counties which are 150 to 250 miles from Raleigh : Ashe, Buncombe, Burke, Caldwell, Camden, Cherokee, Cleveland, Currituck, Haywood, Henderson, Jackson, Macon, Madison, McDowell, Polk, Rutherford, Wetauga, Yancey.

In the first District, Wake county, the sum of annual populations for the five years was 112,129, an annual average of 22,426. These sent 23 patients to the asylum, or 4.6 in each year, which is equal to one in 4,875 of the people yearly.

In the second District, 469,606, the total of annual populations or average of 93,923, sent 73 patients to the asylum in the five years, averaging 14.6, or one in 6,433 of the whole population in each year.

The sum of annual numbers of the people in the third District was 1,708,271, or an average of 341,654 in each year. From among these 176 lunatics were sent to the asylum, which was 35.2, or one in 9,707 of the living in each year.

In the fourth District the total of annual populations was 999,407, averaging 199,881 yearly; 91 lunatics in the five years gave a yearly average of 18.1, or one in 10,982.

In the fifth District the sum of the annual populations was 549,350, or an average of 109,870, who sent 12 patients, or an average of 2.4, which was one in 45,779 in each year.

MISSISSIPPI.

The State Insane Asylum was opened at Jackson, Hinds county, in 1855. The records of the residences of patients for only one year, 1858, are at command.

Hinds county constitutes the first District.

The second District includes the ten counties within 50 miles of the Asylum: Claiborne, Copiah, Leake, Madison, Rankin, Scott, Simpson, Smith, Warren, Yazoo.

The third District includes the thirteen counties 50 to 75 miles of Jackson: Attala, Covington, Franklin, Holmes, Issaqueena, Jasper, Jefferson, Jones, Lawrence, Neshoba, Newton, Washington, Winston.

The fourth District comprises the twenty counties 75 to 125 miles distant: Adams, Amite, Boliver, Carroll, Chickasaw, Choctaw, Clark, Greene, Kemper, Launderdale, Marion, Noxubee, Oktibbeha, Perry, Pike, Sunflower, Tallahatchie, Wayne, Wilkinson, Yallabusha.

D

The fifth District includes, the rest of the State, sixteen counties, 125 to 225 miles distant : Calhoun, Coahoma, De Soto, Hancock, Harrison, Itawamba, Jackson, Lafayette, Lowndes, Marshall, Monroe, Panola, Pontotoc, Tippah, Tishamingo, Tunica.

The first District had 30,036 inhabitants in the year recorded, and sent 2 patients, or one in 15,018.*

The second District had 133,500 people, who sent 19 patients, or one in 7,026.

The third District had 125,018 population, who sent 9 patients to the asylum, or one in 13,890.

The fourth District had 226,123 inhabitants, from whom 14 lunatics went, or one in 16,151.

The fifth District had 276,599, who sent 13 patients, or one in 21.276.

LOUISIANA.

The State Asylum was opened in 1848, in Jackson, Parish of East Feliciana. The record of ten years—November, 1848, to 1858—is available for this report.

East Feliciana constitutes the first District.

The second District embraces nine contiguous parishes or counties within 50 miles of the hospital : Ascension, Avoyelles, East Baton Rouge, Iberville, Livingston, Point Coupée, St. Helena, West Baton Rouge, West Feliciana.

The third District includes fifteen counties, 50 to 100 miles distant : Assumption, Catahoola, Concordia, Jefferson, Lafayette, Madison, Orleans, St. Charles, St. James, St. John Baptist, St. Landry's, St. Martin's, St. Mary's, St. Tammany, Tensas.

* This is the record of a single year only, and is not an exact indication of the permanent habits of the people.

The fourth District includes fifteen counties, 100 to 150 miles distant : Calcasieu, Caldwell, Carroll, Franklin, Jackson, Lafourche, Morehouse, Nachitoches, Plaquemines, Rapides, St. Bernard, Terre Bonne, Vermillion, Washite, Winn.

The fifth District embraces seven counties, 150 to 200 miles from Jackson : Bienville, Bossier, Caddo, Claiborne, De Soto, Sabine, Union.

In the first District the average annual populations was 13,971, who sent 21 patients, or 2.1 per year, or one in 6,653.

The second District had an average annual population of 89,889, who sent 59 lunatics to the asylum, or 5.9 in one year, which was one in 15,235 of the people.

The third District had 164,788 inhabitants, from whom 99 patients went to the asylum. This gave 9.9, or one in 16,645 in each year.

In the fourth District the average annual population was 124,115, from whom 58 patients were sent in the ten years to the asylum; this was 5.8, or one in 21,399 in each year.

In the fifth District the average number of the people was 59,060, who sent 19 patients, or 1.9 yearly to the asylum. This was one in 25,822.*

TENNESSEE.

The Hospital for the Insane was opened at Nashville, Davidson county, and the people of every county invited to send their patients on equal terms. The records

---

* NOTE.—In the reports, a very large and undue proportion of patients are stated to have been sent from the parishes of Orleans and Caddo. These are so large as to create a doubt whether some others besides those, belonging to other places, are not included. These are, therefore, omitted in this classification and statement.

of the residence of the patients from 1852 to 1859 are printed in the reports. The State is divided for the purpose of this report into five districts.

The first District is Davidson county.

The second District includes the six contiguous counties, which are within 35 miles of Nashville : Cheatham, Robertson, Rutherford, Sumner, Williamson, Wilson.

The third District includes nineteen counties, which are from 35 to 70 miles distant : Bedford, Cannon, Coffee, De Kalb, Dickson, Hickman, Humphreys, Jackson, Lewis, Macon, Marshall, Maury, Montgomery, Perry, Putnam, Smith, Stewart, Warren, White.

The fourth District includes thirty-nine counties, 70 to 150 miles from Nashville : Anderson, Benton, Bledsoe, Bradley, Campbell, Carroll, Cumberland, Decatur, Dyer, Fentress, Franklin, Gibson, Giles, Grundy, Hamilton, Hardeman, Hardin, Haywood, Henderson, Henry, Lawrence, Lincoln, McMinn, McNairy, Madison, Marion, Meigs, Monroe, Morgan, Obion, Overton, Polk, Rhea, Roane, Scott, Sequatchie, Van Buren, Wayne, Weakley.

The fifth District comprises nineteen counties, from 150 to 300 miles distant from the hospital : Blount, Carter, Claiborne, Cocke, Fayette, Grainger, Green, Hancock, Hawkins, Jefferson, Johnson, Knox, Lauderdale, Sevier, Sullivan, Tipton, Union, Washington.

In the seven and one-half recorded and published years, the first District had a total sum of annual populations 325,640, who sent 83 patients. The annual averages were 43,418 people and 11 patients, or one in 3,923 of the living.

In the second District the total of the populations for the seven and one-half years was 931,655, who sent

112 lunatics to the hospital. The annual averages were 124,220 people and 14.9 patients, or one in 8,318 of the living, at home.

In the third District the populations were 1,724,574, and the patients 131 during the recorded years. These averaged yearly 229,943 inhabitants and 17.4 lunatics sent to the hospital, or one in 13,164 people.

In the fourth District the sum of the enumerated and calculated populations was 3,147,817, from whom 154 patients went to the hospital. The annual averages of these were 419,708 people and 20.5 lunatics, or one from every 20,440 inhabitants of the district.

In the fifth District the sum of the annual populations through seven and one-half recorded years was 1,582,606, from whom 100 patients went to the hospital. The yearly averages of these were 211,014 people and 13.3 lunatics, or one in 15,826* of the living.

KENTUCKY.

In Kentucky, the asylum at Lexington, Fayette county, was the only institution for the insane in the State, and offered its advantages to the people of every county from 1824 to 1855, when the Western Asylum was opened, and took patients from the western counties until it was burned down in 1860. From that time the Lexington Asylum has received patients from all parts of the commonwealth.

These calculations are based on the experience of the Lexington Asylum from 1824 to 1855, inclusive, except

* Knox county is reported to have sent about three times as many patients as other counties in the same neighborhood in proportion to its population. There may have been an error in the record.

the years 1844, '45, '46 and '47, of which the record has
not been obtained.

Fayette county is the first District.

The second District includes the six contiguous
counties, within 30 miles of Lexington : Bourbon, Clark,
Jessamine, Madison, Scott, Woodford.

The third District includes forty-three counties, 30 to
75 miles from Lexington : Anderson, Bath, Boone,
Boyle, Bracken, Bullitt, Campbell, Carroll, Casey, Estill,
Fleming, Franklin, Gallatin, Garrard, Grant, Greene,
Harrison, Henry, Jackson, Kenton, Lewis, Lincoln,
Marion, Mason, Mercer, Montgomery, Morgan, Nelson,
Nicholas, Oldham, Owen, Owsley, Pendleton, Powell,
Rock Castle, Rowan. Shelby, Spencer, Taylor, Trimble,
Washington, Wolfe.

The fourth District includes thirty-nine counties, 75
to 130 miles from the asylum : Adair, Allen, Barren,
Boyd, Breathitt, Breckenridge, Butler, Carter, Clay,
Clinton, Cumberland, Daviess, Edmondson, Floyd, Gray-
son, Greenup, Hancock, Hardin, Harlan, Hart, Johnson,
Knox, La Rue, Laurel, Lawrence, Letcher, Logan,
Meade, Metcalf, Monroe, Ohio, Perry, Pike, Pulaski,
Russell, Simpson, Warren, Wayne, Whitley.

The fifth District comprises the twenty-one counties
which are from 130 to 300 miles from Lexington : Bal-
lard, Caldwell, Callaway, Christian, Crittenden, Fulton,
Graves, Henderson, Hickman, Hopkins, Livingston,
Lyon, MacCracken, MacLean, Magoffin, Marshall, Muh-
lenburgh, Todd, Trigg, Union, Webster.

In the first District the sum of populations was
574,655, averaging 21,283. These sent 180 patients,
or an annual average of 6.6, or one in 3,198 of the
people.

In the second District the sum of annual populations was 2,134,144, or an average of 79,042, who sent 200 patients, equal to 7.4, or one in 10,670 of the people yearly.

In the third District the sum of annual numbers of the people was 7,908,111, or 292,892 in each year. These sent 610 patients, equal to 22.5, or one in 12,964 of the people yearly.

In the fourth District the sum of populations was 6,250,198, or 231,488 in each year. From these 259 patients went to the asylum, which was equal to 9.5, or one in 24,132 of the annual number living.

In the fifth District the total population of the twenty-eight years amounted to 3,058,111, or 113,263 yearly. From among these 110 lunatics were sent to the asylum, which was an annual average of nearly 4, or one in 27,801 people.

MISSOURI.

In Missouri the Asylum at Fulton, Callaway county, was opened in 1851. It was suspended during the years 1861, 1862 and 1863, and again reöpened. The record of the residence of the patients received during these eleven years is printed, and forms the basis of the calculations.

The first District consists of Callaway county.

The second District includes the six contiguous counties, within 50 miles: Audrain, Boone, Cole, Gasconade, Montgomery, Osage.

The third District includes twenty-one counties, from 50 to 75 miles from Fulton: Cooper, Crawford, Franklin, Howard, Lincoln, Macon, Maries, Marion, Miller, Moniteau, Morgan, Pettis, Phelps, Pike, Pulaski, Ralls, Randolph, Saline, Shelby, St. Charles, Warren.

The fourth District includes twenty-six counties, from 75 to 125 miles from Fulton : Adair, Benton, Camden, Carroll, Chariton, Clark, Dallas, Dent, Henry, Hickory, Iron, Jefferson, Johnson, Knox, Laclede, Lafayette, Lewis, Linn, Livingston, Monroe, St. Francis, Ste. Genevieve, Scotland, Texas, Washington, Wright.

The fifth District includes fifty-eight counties, from 125 to 225 miles distant from the asylum: Andrew, Atchison, Barry, Barton, Bates, Bollinger, Buchanan, Butler, Caldwell, Cape Girardeau, Carter, Cass, Cedar, Christian, Clay, Clinton, Dade, Daviess, De Kalb, Douglas, Dunklin, Gentry, Greene, Grundy, Harrison, Holt, Howell, Jackson, Jasper, Lawrence, McDonald, Madison, Mercer, Mississippi, New Madrid, Newton, Nodaway, Oregon, Ozark, Pemiscot, Perry, Platte, Polk, Putnam, Ray, Reynolds, Ripley, St. Clair, Schuyler, Scott, Shannon, Stoddard, Stone, Sullivan, Taney, Vernon, Wayne, Webster.

In the first District the sum of the populations through eleven years was 147,751, who sent 25 patients. The annual average was, of population 13,432, and of patients 2.3, or one in 5,910 people.

In the second District the total sum of populations was 590,009. These sent 47 patients. The annual averages were, of population 53,637, and of patients 4.3, or one in 12,553 of the people.

In the third District the sum of the annual populations was 2,168,390, who sent 155 patients to the asylum. The averages of each year were, of population 197,126, and of patients 14.1, or one in 13,989 of the living at home.

In the fourth District the sum of the populations of all the eleven years was 1,934,221, from whom 121 patients went to the asylum. The averages of the several years were, of population 175,838, and of patients 11, or one in 15,983.

In the fifth District the sum of population of the several years was 3,905,390. From these 147 patients were sent to the hospital. The annual averages were, of population 355,035, and of patients 13.2, or one in 26,933 of the people.

CANADA WEST.

The Provincial Hospital of Canada West is in Toronto, York county. It was opened in 1853, and the record of the residences of the patients stated in the reports from that time to 1865, twelve years. The hospital has been offered equally to all the people of the Province.

The county of York constitutes the first District.

The second District includes three contiguous counties : Halton, Ontario, Peel.

The third District includes eleven counties, within 35 to 70 miles : Brant, Durham, Haldimand, Lincoln Simcoe, Victoria, Waterloo, Welland, Wellington, Wentworth, Hamilton City.

The fourth District includes fifteen counties, 70 to 150 miles from Toronto : Addington, Bruce, Elgin, Grey, Hastings, Huron, Lenox, London City, Middlesex, Norfolk, Northumberland, Oxford, Perth, Peterboro, Prince Edward.

The fifth District includes eighteen counties, 150 to 300 miles distant : Algona District, Ottawa City, Carleton, Dundas, Essex, Frontenac, Kingston, Glengary, Grenville, Kent, Lambton, Lanark, Leeds, Nipissing, Prescott, Renfrew, Russell, Stormont.

E

In the first District the sum of the annual populations for the twelve years under observation was 1,251,201, or an average of 104,266. They sent 393 patients, or 32.7, which was one in 3,183 people in each year.

In the second District the sum of people through the whole period was 1,105,797, an annual average of 92,149. From these 153 patients went to the hospital, equal to 12.7, or one in 7,227 of the people yearly.

In the third District the sum of populations was 4,181,592, an average of 348,466. The whole number of their patients in the hospital was 540, averaging 45, or one in 7,743 of the people yearly.

In the fourth District the sum of annual populations was 5,598,521, being an average of 466,543. Their patients in the hospital were 444 during the whole period, equal to 37, or one in 12,608 yearly.

In the fifth District the whole sum of populations was 4,331,015, an average of 360,917 yearly. They sent 297 patients in the whole period, or 24.7, equal to one in 14,582 people yearly.

NOVA SCOTIA.

The Provincial Hospital was opened in 1858, for the equal use of all the people of the Province. The records of seven years have been printed, showing the residence of the patients who were received from 1858 to 1864, inclusive.

The population of 1860 only has been obtained. No calculation is therefore made of that of the other years, but as this was near the middle of the period it will be, at least, near the truth, to assume this as the average of each of the years of the hospital operations that are known.

The Province is, for the purposes of this report, divided into four Districts.

Halifax county is the first District.

The second District includes four contiguous counties, within 65 miles of the hospital: Colchester, Hants, Lunenburgh, Pictou.

The third District includes six counties, from 65 to 100 miles from the hospital: Annapolis, Cumberland, Guysborough, Kings, Queens, Sidney.

The fourth District includes seven counties, from 100 to 175 miles from Halifax: Cape Breton, Digby, Inverness, Richmond, Shelburne, Victoria, Yarmouth.

In the first District the population was 49,021 in 1860. 105 patients went to the hospital in the seven years, or a yearly average of 15, equal to one in 3,268 of the people.

In the second District there were, in 1860, 85,922 people, who sent 84 patients in the seven years, an annual average of 12, or one in 7,160 persons living.

In the third District the population was 91,966 in 1860, who sent 52 patients to the hospital, an average of 7.4, or one in 12,427 people in each year.

In the fourth District the number of the people in 1860 was 103,948, and their patients in the hospital were 34 during the seven years. This is equal to an annual average of 4.85, or one in 21,432 of the people.

Population to One Patient Annually Sent to Lunatic Hospitals.

STATE.	Number of Years.	DISTRICTS.				
		I.	II.	III.	IV.	V.
Maine,.............	1840–65	2,835	5,171	5,630	7,890	....
New Hampshire,...	1842–65	2,440	3,470	6,280	....	....
Massachusetts,.....	1833–53	2,229	3,872	4,953	....	....
Rhode Island,......	1849–65	3,094	5,279	....	....	....
New York,........	1843–65	2,772	5,820	7,351	11,535	....
New Jersey,.......	1848–66	2,253	3,714	5,905	....	...
Pennsylvania,......	1850–57	6,061	10,793	17,686	23,748	...
East Pennsylvania,.	1857–66	5,884	10,497	17,414	53,629	....
West Pennsylvania,	1857–66	3,650	10,585	22,382	....	....
Maryland,.........	1850–64	7,034	10,122	23,009	....	....
Virginia,..........	1828–59	5,472	12,314	21,570	24,433	25,105
North Carolina,....	1856–60	4,875	6,433	9,707	10,982	45,779
Mississippi,........	1858	*15,018	7,026	13,890	16,151	21,276
Louisiana,.........	1848–58	6,653	15,235	16,645	21,399	25,822
Tennessee,.........	1852–59	3,923	8,318	13,164	20,440	*15,826
Kentucky, ...	†1824–55	3,198	10,670	12,964	24,132	'27,801
Ohio,.............	1838–66	5,060	7,304	11,712	28,873	....
Illinois, ...	1847–64	3,306	7,865	9,317	11,753	15,585
Michigan,.........	1859–65	3,162	9,229	11,089	14,208	58,039
Missouri,..........	‡1851–64	5,910	12,553	13,989	15,983	26,933
Canada,...........	1853–66	3,184	7,227	7,744	12,608	14,582
Nova Scotia,......	1858–64	467	1,023	1,768	3,057	....

* There is apparently something unexplained in the record of one county in each of these Districts.

† Excluding 1844, '45, '46 and '47.   ‡ Excluding 1861, '62 and '63.

In all these States the privileges of the hospitals are offered equally to the people of the counties. The patients of Oneida and Allegany counties in New York, of Mercer and Warren counties in New Jersey, of Dauphin and Venango counties in Pennsylvania, can enter on the same terms, enjoy the same advantages, and for the same price. The only difference is the burden of cost, care and labor of travel from their homes to the place of healing. And yet the actual use of the hospitals by, and the practical value of these institutions to the people of the remote districts have been only one-fourth as great in New York, about one-third as great in New Jersey, and less than one-third as great in Pennsylvania as they have been in the districts near to them.

Similar discrepancies in favor of the central counties and against the distant counties are seen to have existed in all the other States whose record has been obtained.

EFFECT OF MULTIPLYING HOSPITALS IN STATES.

This principle has been remarkably manifested whenever and wherever a second hospital has been opened in any State and placed in a district remote from the one previously in operation. The people who sent a few patients to the distant institution, now sent many to the hospital which was brought to their neighborhood. The number of lunatics that found a place of healing was suddenly and permanently increased.

In MASSACHUSETTS, the Hospital at Worcester was the only State institution for the insane in the commonwealth from 1833 to 1854, when the second hospital was opened in Taunton, Bristol county, for the southeastern part of the State. The Worcester establishment continued to receive all the patients from the northern, central and western counties until 1858, when the third hospital was opened in Northampton, Hampshire county, for the western district. In both of these districts there was a sudden and large increase of the insane, whose friends sought and used these new places of healing for them.

During the eight years, 1845 to 1853, previous to the opening of the Taunton Hospital, the people of Bristol county had sent 151 patients to Worcester, which was an annual average of one patient in 4,434 inhabitants.

During the eight years after the hospital was opened within their borders they sent 324 patients to it, which was an annual average of one patient in 2,194 people.

In the former period the people of Plymouth county sent one in 3,719 of their number, and in the latter period one in 2,774.

Barnstable, Dukes and Nantucket counties sent, in the former period, one in 4,118, and in the latter, one in 3,573 to the hospitals.

Population for One Patient sent Annually to the State Hospitals.

COUNTY.	1845 TO 1853.			1854 TO 1862.			RATE OF INCREASE.
	Patients.	Sum of Population.	People to 1 Patient.	Patients.	Sum of Population.	People to 1 Patient.	
Bristol,......	151	669,581	4,434	324	810,903	2,194	102.1
Plymouth,....	132	493,215	3,719	204	565,981	2,774	34.
Barnstable, Nantucket, Dukes,....	104	429,319	4,118	118	421,662	3,573	15.2
Five Counties	387	1,592,115	4,111	646	1,798,546	2,784	42.9

During the four years—1854 to 1858—the people of Hampshire county sent 37 patients to the Worcester Hospital, which was an annual average of one in 4,008 inhabitants. In the four years after the opening of the third hospital in their midst, the same people sent 85 people, or one in 1,787 of their number to its care.

Franklin county sent in the former period 19 patients, or one in 6,574 people to Worcester, and in the latter period 52, or one in 2,419 people to Northampton.

Berkshire county is geographically fifty miles nearer to Northampton than to Worcester. But a range of mountains lies between, and the roads are difficult for travellers, who can use only private conveyances, except the Western Railroad to Springfield, and the Connecticut River Railroad from Springfield to Northampton. This practically reduces the difference of distance between the two hospitals to thirty miles. And many when once in the cars on the Western Road, find it easier to continue fifty-four miles further to Worcester than to change cars and go twenty miles to Northampton with their patients. Therefore the increase is less in Berkshire county than in the others. Nevertheless, there was an increase.

Before 1858 the Berkshire people sent 33 patients, or one in 6,937 people yearly to Worcester, and after that they sent to Worcester and Northampton 47 patients, or an average in each year of one in 4,715 people.

To the towns in the eastern part of Hampden county Worcester is nearer and more accessible than Northampton. Most of the people must necessarily use the Western Railroad, whether going to Worcester or Northampton, and all must change cars at Springfield if they go to Northampton, but not if they go to Worcester.

The people of Hampden county sent in the former period one in 2,185 of the living to Worcester, and in the latter, one in 1,988 in each year.

Population to One Patient sent to Hospital before and after Northampton Hospital was opened, Western District.

COUNTY.	1855 to 1858, four years.			1859 to 1862, four years.			Increase.
	Patients sent.	Sum of annual Populations.	People to 1 Patient.	Patients sent.	Sum of annual Populations.	People to 1 Patient.	Per ct. Patients sent
Berkshire,...	33	212,437	6,437	47	221,640	4,715	38.6
Franklin.....	19	124,916	6,574	52	125,830	2,419	171.2
Hampshire,..	37	148,294	4,008	85	151,897	1,787	124.3
Hampden,...	101	220,680	2,185	116	230,784	1,988	9.9
Four Counties	190	706,327	3,717	300	730,151	2,433	52.7

The people of Hampshire county nearly trebled the number and proportion of their patients in the hospital; the people of Franklin and Bristol more than doubled them, and the other counties also increased them very greatly, and thus so many more of their lunatics found places of healing and protection when the hospital was brought to their neighborhood and within their reach.

In Ohio the State Hospital at Columbus received patients from all the counties from 1838 to 1858, when the Northern Asylum at Newburgh, Cuyahoga county, and the Southern Asylum at Dayton, Montgomery county, went into operation and received patients from certain surrounding districts, which were defined by the law.

The northern districts had sent to Columbus 403 patients, an annual average of one patient in 13,201 of the population during the twelve years previous to the opening of the hospital in their midst at Newburgh.

During the next three years and eight months after the new hospital was opened, the same people sent to its care 549 patients, an annual average of one patient in every 3,138 of their number living during these years.

In the first period of twelve years, 1838 to 1850, the people of the Southwestern District had sent to the asylum at Columbus 373 patients, which is an annual average of one in 13,126 people. During the nine years and two months next after the new asylum was opened in their own neighborhood, at Dayton, they sent 1,079 patients to its care, which is an annual average of one in 4,688 people living in each of these years.

Population to One Patient sent.

District.	To Columbus.	Home Hospital.	Increase Per Cent. o Patients.
Northern,.............	13,201	3,138	420
Southern,............	13,128	4,304	305

In Kentucky, from the opening of the Lunatic Hospital at Lexington in 1824, to the end of 1855, it was the only institution for the insane in the State. In 1855, the second hospital was opened at Hopkinsville for the patients in the western part of the State.

The published records do not furnish means of determining the number of patients sent to the Western Hospital from each county through each of the years from 1855 to 1860, when it was burned; but, comparing the reports of this institution with the Eastern, it is found that the twenty-six most westerly counties, within 80 miles of Hopkinsville, sent through the twenty-eight years, 1824 to 1855, (excepting 1844 to 1847, of which no records are to be found,) 203 patients, which is an annual average of 7.2, or one in 15,015 people. During the three years after the new hospital was opened in their midst, they sent 117 patients, which was 39, or one in 6,271 of the people in each year.

During the twenty-eight years, the whole State sent 1,543 patients, or one in 12,913 people annually to their single central hospital. During the four years next following the opening of the Western Hospital, the whole State sent 513 patients to their two hospitals, or one annually in 8,017 of the population.

EFFECT OF RAILROADS AND OTHER FACILITIES OF TRAVEL.

Facilities of travel, navigable rivers, canals, railroads, public highways, public conveyances, which render communication easy and cheap, and intercourse familiar, and virtually diminish distance from the hospital, increase the ratio of patients that are sent to it. We therefore find that those counties which are situated along the course of rivers, canals, roads, etc., leading directly to the situation of the hospitals, have sent more patients to these institutions than other counties of equal population

F

and at equal distances, but not favored with these facilities of communication.

Ten counties in New York along the line of the railroad, canal, etc., east and west of Utica, with easy means of travel, having a sum of annual populations equal to 15,622,250, sent 2,151 patients to Utica; while, during the same period, ten other counties, northeast and southwest from Utica, with no easy means of communication, with a sum of 7,840,684 annual populations, sent 647 patients, or one in 11,934 of their number to the State Hospital.

Taking all these facts into view, we have here indisputable proof of the effect of distance in diminishing the practical benefits of lunatic hospitals to the people of any district. In all these States these hospitals are as open and their advantages as freely granted to the patients from the most remote towns as to those in their very neighborhood. It is not hinted, or even suspected, that the lunatics whose friends reside afar off are not as kindly, as faithfully, and as successfully treated, and at as small a cost as those whose friends are so near as to keep a watchful vigilance over their welfare.

A HOSPITAL IS BETTER KNOWN TO THE NEIGHBORING PEOPLE.

The idea of the hospital purposes and its management is familiar to those who live in its vicinity. They know its means, its objects, and its administration; they know the character of its officers and its attendants.

They are frequently witnessing its operations and results in the many who are going to and returning from it, in improved or restored mental health. Whenever they think of the possibility of their becoming insane, the idea of the hospital presents itself to their minds, in the same connection, almost as readily as the idea of their own chambers, their own physician, and the tender nursing of their own family is associated with the thought of having a fever or dysentery. And, when any one of their family or friends becomes deranged, the hospital occurs to them as a means of relief, and they look upon it as a resting place from their troubles.

But this ready association of the hospital with lunacy, and this generous confidence in its management diminishes as we recede from it. The people in the remoter places know the general facts, but distance lends an obscurity to the notion, and thus the character of the hospital and its administration do not stand before them, as the thought of home and domestic arrangements, of which they can cheerfully and trustfully avail themselves in any emergency. To them the hospital seems a strange place—perhaps a place of unkind restraint or even of needless confinement, rather than a home of tenderness. Its officers are to them strangers rather than friends; and its attendants, though good and honest men, are not as household comforters and nurses, or even as neighbors, whose ready and affectionate sympathy is sure, and on whom they are accustomed to call in time of trouble, and to whom they unhesitatingly commit the care of their disordered and distressed relatives or children.

Then the unwillingness to be far separated from their suffering or weakened friends operates with many. This is indeed a mere feeling or sentiment; but it is converted into practical facts, and retains some at home who would otherwise be sent to and cured in a hospital if it were nearer to them. The State Lunatic Hospital, when it is used, is no better to the people of Oneida than to those of Chautauqua, Cattaraugus and Clinton; but so long as a portion of the people of the remote counties do not feel so, their insane friends are not sent there.

The difficulties and expense of sending lunatics over long distances, or unfrequented and indirect roads, or by private conveyances, are perhaps the most effectual obstacles in the way, and more than any other diminish the number of patients, with the increase of miles, that separate them from the hospital.

For these reasons the towns in the neighborhood of the public hospital in this State have enjoyed more than four times as much of its benefits as the remote towns ; and all the other hospitals mentioned in this article have been compelled to confer their blessings in a similar, and some of them in a much greater disproportion upon the people of the neighboring than upon those of the distant districts of the States to which they respectively belong.

We think we have here presented facts enough to establish it as a general principle, that the advantages of any public lunatic hospital, however freely and equally they may be offered to all the people of any State, are yet, to a certain degree, local in their operation, and are enjoyed by people and communities to an extent in proportion to their nearness to or distance from it.

Whenever and wherever the same causes exist, the same effects must be produced, and any hospital that may be hereafter established must be subject to the same law.

This law of nearness, inviting and increasing the patients, and of distance, preventing them and diminishing the number in hospital, is our very nature, and must operate in the future as well as the past. The people will be influenced by the same motives in time to come as they have been in the years that have gone by.

There are then two policies in regard to providing for the insane presented to the people of New York for their adoption. One is to continue the present plan of having all the patients sent to a central establishment from all the State; the other proposes to create small local asylums in the west, northeast and southeast, in the midst of the people who wish to use them for their insane.

The unequal and unjust operation of the former plan has been demonstrated in this report, and is manifestly a necessary and natural law which will operate in all places and all times, in similar circumstances. The other plan will give to all parts of the State the privileges which have been hitherto withheld from them, but which have been enjoyed by the central counties, of easy and frequent access to the means of healing their insane, and of a larger proportion of them thus restored to health.

Whatever is now done, whether it be the building of the new and great central establishment for the incurable, or the creation of the smaller local institutions, it will be the plan for twenty years, or more, to come; for another institution with that already now in operation, will seem to meet all the wants of the State, and to accommodate all who seem to need it, or all who are within accessible distance, for the next, as the Utica Asylum has for the last twenty years.

Yet the same inequality will remain. The many will be sent from the neighboring counties, and all the recent cases in Oneida county, and all the old cases in the county of the incurable and chronic asylum, and nearly all of those in contiguous counties will be thus provided for; while a few, varying from one-quarter to one-half as large a proportion will be sent from the remote and remotest districts, and the remainder, the majority of those attacked and of those needing the healing and soothing and protecting influence of the asylum, will be left at home, without means of restoration or proper guardianship.

On the other hand, if the other policy is adopted, and asylums are established in the west, northeast and southeast districts, in the midst of and accessible to the patients, those districts will send as large a proportion of their insane to be healed and to be cared for as are now and have been sent from Oneida and the neighboring counties to Utica.

It is then for the Legislature to decide, and especially for the representatives of the counties 100 miles and more from Utica, whether this unequal provision shall be continued; whether the bounties of the State shall be so liberally given to the central portion and so sparingly allowed to the remote parts of the State.